Houghton Mifflin Harcourt

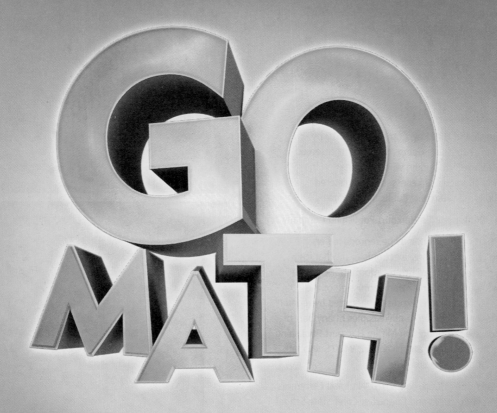

D1275578

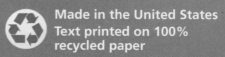

Made in the United States
Text printed on 100% recycled paper

Houghton Mifflin Harcourt

Printed in the U.S.A.

ISBN 978-0-544-34169-2

19 0928 20

4500800109   B C D E F G

Dear Students and Families,

Welcome to **Go Math!**, Grade K! In this exciting mathematics program, there are hands-on activities to do and real-world problems to solve. Best of all, you will write your ideas and answers right in your book. In **Go Math!**, writing and drawing on the pages helps you think deeply about what you are learning, and you will really understand math!

By the way, all of the pages in your **Go Math!** book are made using recycled paper. We wanted you to know that you can Go Green with **Go Math!**

Sincerely,

The Authors

Made in the United States
Text printed on 100% recycled paper

Authors

Juli K. Dixon, Ph.D.
Professor, Mathematics Education
University of Central Florida
Orlando, Florida

Edward B. Burger, Ph.D.
President, Southwestern University
Georgetown, Texas

Steven J. Leinwand
Principal Research Analyst
American Institutes for
   Research (AIR)
Washington, D.C.

Contributor

Rena Petrello
Professor, Mathematics
Moorpark College
Moorpark, California

Matthew R. Larson, Ph.D.
K-12 Curriculum Specialist for
   Mathematics
Lincoln Public Schools
Lincoln, Nebraska

Martha E. Sandoval-Martinez
Math Instructor
El Camino College
Torrance, California

English Language Learners Consultant

Elizabeth Jiménez
CEO, GEMAS Consulting
Professional Expert on English
   Learner Education
Bilingual Education and
   Dual Language
Pomona, California

Number and Operations

Critical Area Representing, relating, and operating on whole numbers, initially with sets of objects.

Critical Area

GO DIGITAL

Go online! Your math lessons are interactive. Use *iTools*, Animated Math Models, the Multimedia *e*Glossary, and more.

Chapter 2 Overview

In this chapter, you will explore and discover answers to the following **Essential Questions**:

- How can building and comparing sets help you compare numbers?
- How does matching help you compare sets?
- How does counting help you compare sets?
- How do you know if the number of counters in one set is the same as, greater than, or less than the number of counters in another set?

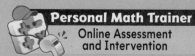

Personal Math Trainer
Online Assessment and Intervention

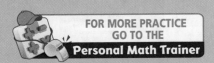

FOR MORE PRACTICE
GO TO THE
Personal Math Trainer

Practice and Homework

Lesson Check and
Spiral Review in
every lesson

Compare Numbers to 5

Curious About Math with Curious George

Butterflies have taste buds in their feet so they stand on their food to taste it!

- Are there more butterflies or more flowers in this picture?

Name _____

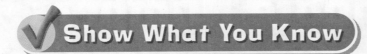
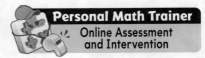
One-to-One Correspondence

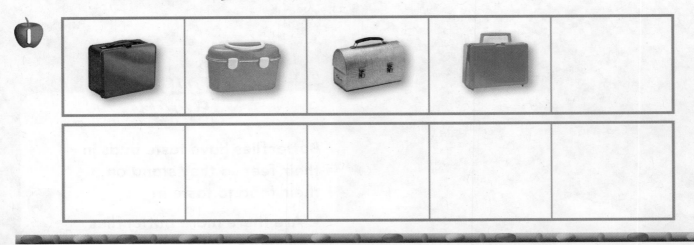

Model Numbers 0 to 5

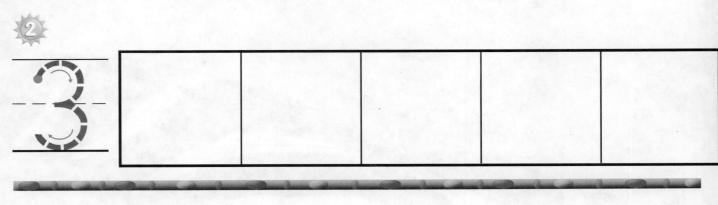

Write Numbers 0 to 5

This page checks understanding of important skills needed for success in Chapter 2.

DIRECTIONS I. Draw one apple for each lunch box. **2.** Place counters in the five frame to model the number. Draw the counters. Trace the number. **3–4.** Count and tell how many. Write the number.

Vocabulary Builder

four

three

three

one

two

five

DIRECTIONS Circle the sets with the same number of animals. Count and tell how many trees. Draw a line below the word for the number of trees.

• **Interactive Student Edition**
• **Multimedia *eGlossary***

Game

Counting to Blastoff

Player 1

5	4	3	2	1	0

Player 2

5	4	3	2	1	0

DIRECTIONS Each player tosses the number cube and finds that number on his or her board. The player covers the number with a counter. Players take turns in this way until they have covered all of the numbers on the board. Then they are ready for blastoff.

MATERIALS 6 counters for each player, number cube (0–5)

Chapter 2 Vocabulary

compare

comparar

13

fewer

menos

23

greater

mayor

31

less

menor, menos

38

match

emparejar

41

more

más

43

one

uno

47

same number

el mismo número

57

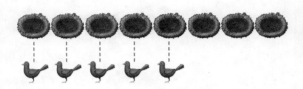

3 **fewer** birds

The number of blue counters **compares** equally to the number of red counters.

3 is **less** than 4

9 is **greater** than 6

2 **more** leaves

Each counter has a **match**.

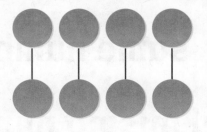

same number of red counters in each row

1

Bingo

© Houghton Mifflin Harcourt Publishing Company

Word Box
- compare
- fewer
- greater
- less
- match
- more
- one
- same number

Player 1

fewer	same number	match	more	greater	less

Player 2

greater	more	one	fewer	compare	same number

DIRECTIONS Shuffle the Vocabulary Cards and place them in a pile. A player takes the top card and tells what he or she knows about the word. The player puts a counter on that word on the board. Players take turns. The first player to cover all the words on his or her board says "Bingo."

MATERIALS 2 sets of Vocabulary Cards, 6 two-color counters for each player

The Write Way

DIRECTIONS Draw to show how to compare sets.
Reflect Be ready to tell about your drawing.

Name _____

Same Number

Essential Question How can you use matching and counting to compare sets with the same number of objects?

Common Core **Counting and Cardinality—K.CC.C.6**
Also K.CC.B.4b, K.CC.C.7
MATHEMATICAL PRACTICES
MP3, MP5

 Listen and Draw **Real World**   Hands On

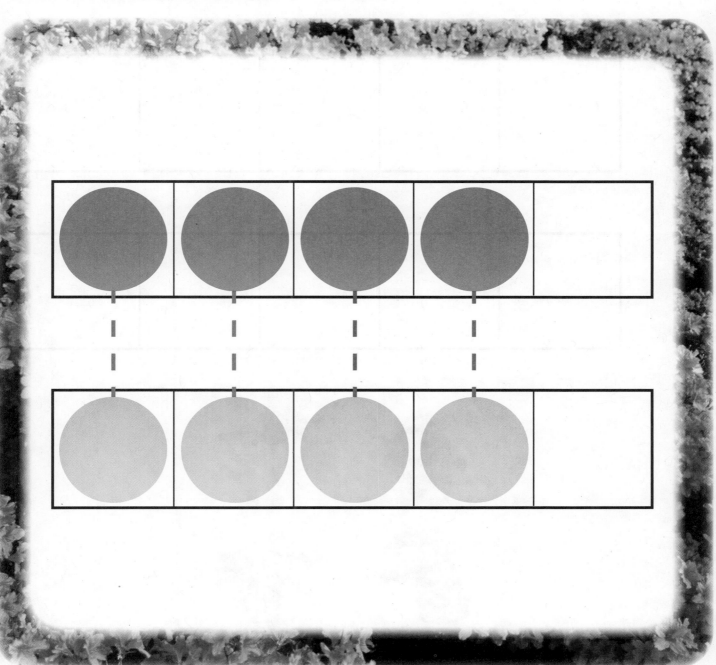

DIRECTIONS Place counters as shown. Trace the lines to match each counter in the top five frame to a counter below it in the bottom five frame. Count how many in each set. Tell a friend about the number of counters in each set.

Chapter 2 • Lesson 1

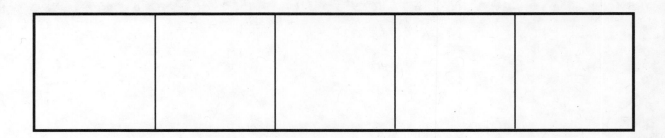

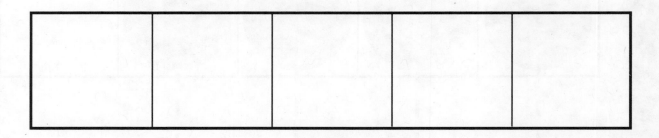

DIRECTIONS 1. Place a counter on each car in the set as you count them. Move the counters to the five frame below the cars. Draw the counters. Place a counter on each finger puppet in the set as you count them. Move the counters to the five frame above the puppets. Draw those counters. Is the number of objects in one set greater than, less than, or the same as the number of objects in the other set? Draw a line to match a counter in each set.

_____

- - - - - - - - -

_____

_____

- - - - - - - - -

_____

DIRECTIONS **2.** Compare the sets of objects. Is the number of hats greater than, less than, or the same as the number of juice boxes? Count how many hats. Write the number. Count how many juice boxes. Write the number. Tell a friend what you know about the number of objects in each set.

Problem Solving • Applications Real World

3

_____

- - - - - - - -

_____

- - - - - - - -

_____

4

DIRECTIONS 3. Count how many buses. Write the number. Draw to show a set of counters that has the same number as the set of buses. Write the number. Draw a line to match the objects in each set. **4.** Draw two sets that have the same number of objects shown in different ways. Tell a friend about your drawing.

HOME ACTIVITY • Show your child two sets that have the same number of up to five objects. Have him or her identify whether the number of objects in one set is greater than, less than, or has the same number of objects as the other set.

84 eighty-four

Name _____

Greater Than

Essential Question How can you compare sets when the number of objects in one set is greater than the number of objects in the other set?

Listen and Draw

Common Core **Counting and Cardinality—K.CC.C.6** *Also K.CC.C.7*
MATHEMATICAL PRACTICES
MP2, MP3, MP5

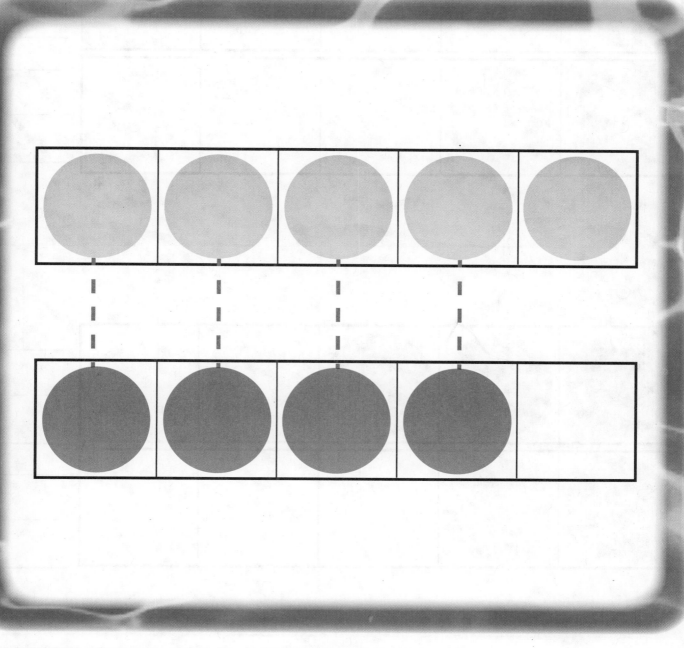

DIRECTIONS Place counters as shown. Trace the lines to match a counter in the top five frame to a counter below it in the bottom five frame. Count how many in each set. Tell a friend which set has a number of objects greater than the other set.

1

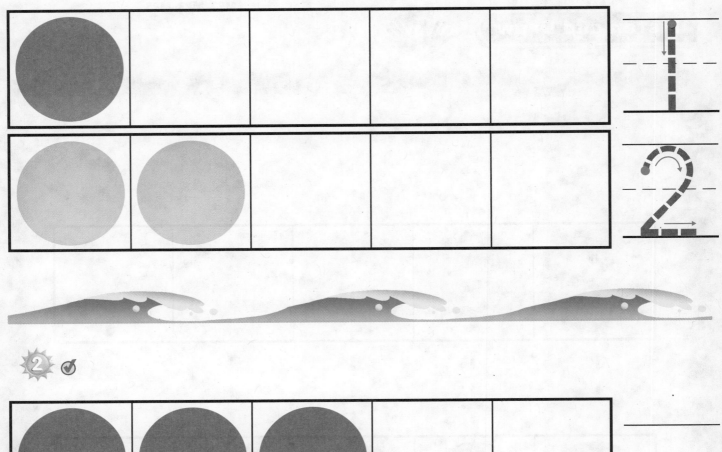

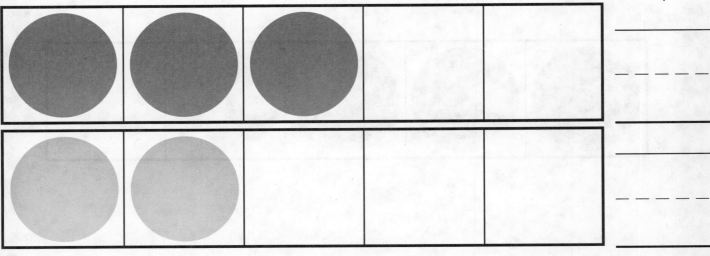

DIRECTIONS 1. Place counters as shown. Count and tell how many in each set. Trace the numbers. Compare the sets by matching. Circle the number that is greater. 2. Place counters as shown. Count and tell how many in each set. Write the numbers. Compare the sets by matching. Circle the number that is greater.

3

_____

- - - - - - -

_____

_____

- - - - - - -

_____

4

_____

- - - - - - -

_____

_____

- - - - - - -

_____

DIRECTIONS 3–4. Place counters as shown. Count and tell how many in each set. Write the numbers. Compare the numbers. Circle the number that is greater.

Problem Solving • Applications

5

DIRECTIONS   5. Brianna has a bag with three apples in it. Her friend has a bag with a number of apples that is one greater. Draw the bags. Write the numbers on the bags to show how many apples. Tell a friend what you know about the numbers.

HOME ACTIVITY • Show your child a set of up to four objects. Have him or her show a set with a number of objects greater than your set.

Name _____

Less Than

Essential Question How can you compare sets when the number of objects in one set is less than the number of objects in the other set?

HANDS ON Lesson 2.3

Counting and Cardinality—K.CC.C.6
Also K.CC.C.7
MATHEMATICAL PRACTICES
MP2, MP3, MP5

Listen and Draw

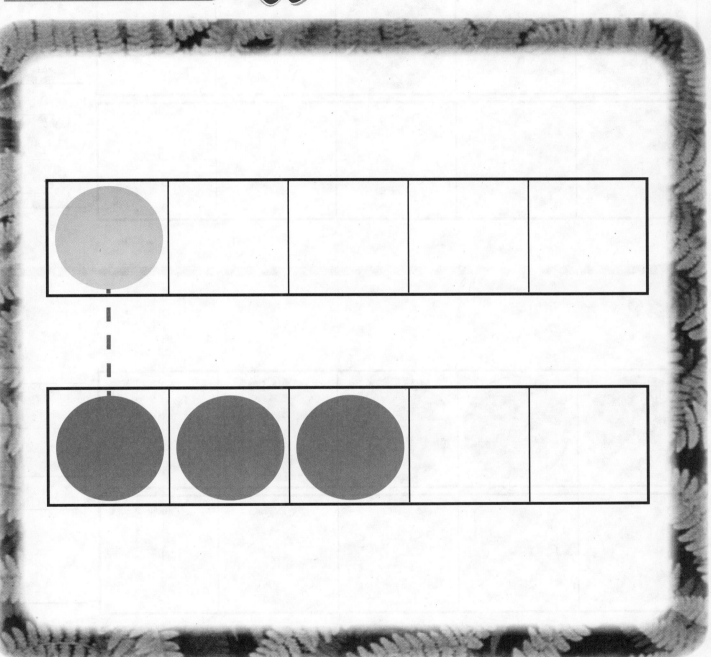

DIRECTIONS Place counters as shown. Trace the line to match a counter in the top five frame to a counter below it in the bottom five frame. Count how many in each set. Tell a friend which set has a number of objects less than the other set.

Chapter 2 • Lesson 3

ninety-three **93**

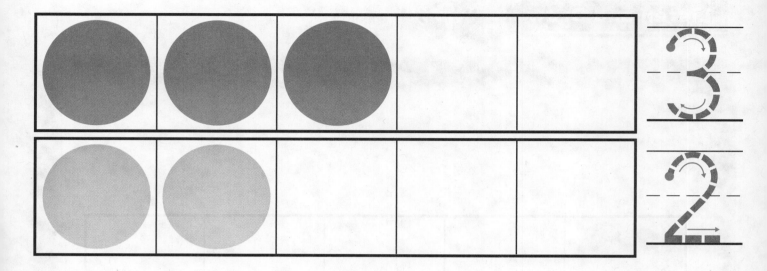

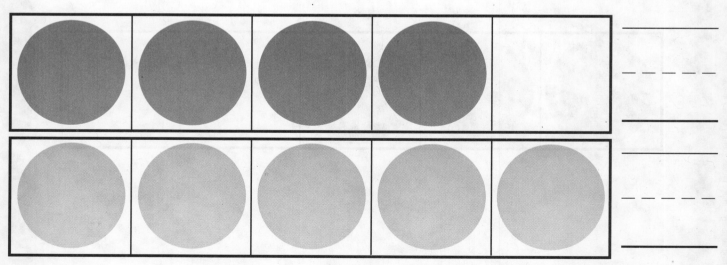

DIRECTIONS 1. Place counters as shown. Count and tell how many in each set. Trace the numbers. Compare the sets by matching. Circle the number that is less. 2. Count and tell how many in each set. Write the numbers. Compare the sets by matching. Circle the number that is less.

Name _____

3

_____

– – – – –

_____

●●

4

_____   _____

– – – – –   – – – – –

_____   _____

●●

DIRECTIONS 3–4. Count and tell how many in each set. Write the numbers. Compare the numbers. Circle the number that is less.

HOME ACTIVITY • Show your child a set of two to five objects. Have him or her show a set of objects that has a number of objects less than you have.

Mid-Chapter Checkpoint

Concepts and Skills

_____

- - - - - - - - -

_____

- - - - - - - - -

_____

- - - - - - - - -

_____

    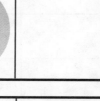

- - - - - - - - -

_____

 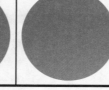  

- - - - - - - - -

_____

THINK SMARTER

DIRECTIONS 1. Place a counter below each object to show the same number of objects. Draw and color each counter. Write how many objects in each row. (K.CC.C.6) **2.** Place counters as shown. Count and tell how many in each set. Write the numbers. Compare the sets by matching. Circle the number that is greater. (K.CC.C.6) **3.** Count the fish in the bowl at the beginning of the row. Circle all the bowls that have a number of fish less than the bowl at the beginning of the row. (K.CC.C.6)

Name _____

Problem Solving • Compare by Matching Sets to 5

Essential Question How can you make a model to solve problems using a matching strategy?

Common Core **Counting and Cardinality—K.CC.C.6**
Also K.CC.C.7
MATHEMATICAL PRACTICES
MP3, MP4, MP5

Unlock the Problem

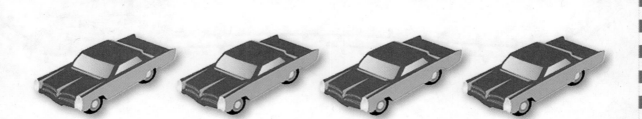

DIRECTIONS These are Brandon's toy cars. How many toy cars does Brandon have? Jay has a number of toy cars that is less than the number of toy cars Brandon has. Use cubes to show how many toy cars Jay might have. Draw the cubes. Use matching to compare the sets.

Chapter 2 • Lesson 4

Try Another Problem

1

2

3 ✓

DIRECTIONS **1.** How many cubes? Trace the number. **2–3.** Model a cube train that has a number of cubes greater than 3. Draw the cube train. Write how many. Use matching to compare the cube trains you drew. Tell a friend about the cube trains.

4

_____

— — — — — — —

_____

5

_____

— — — — — — —

_____

6

_____

— — — — — — —

_____

DIRECTIONS **4.** How many cubes? Write the number.   **5–6.** Model a cube train that has a number of cubes less than 5. Draw the cube train. Write how many. Use matching to compare the cube trains you drew. Tell a friend about the cube trains.

On Your Own

WRITE Math

7

8

_____

_____

_____

_____

_____

_____

_____

_____

_____

_____

DIRECTIONS   7. Kendall has a set of three pencils. Her friend has a set with the same number of pencils. Draw to show the sets of pencils. Compare the sets by matching. Write how many in each set.   **8.** Draw to show what you know about matching to compare two sets of objects. Write how many in each set.

HOME ACTIVITY • Show your child two sets with a different number of objects in each set. Have him or her use matching to compare the sets.

Name _____

Compare by Counting Sets to 5

Essential Question How can you use a counting strategy to compare sets of objects?

Common Core **Counting and Cardinality—K.CC.C.6** *Also K.CC.C.7*

MATHEMATICAL PRACTICES
MP2, MP3, MP6

Listen and Draw (Real World)

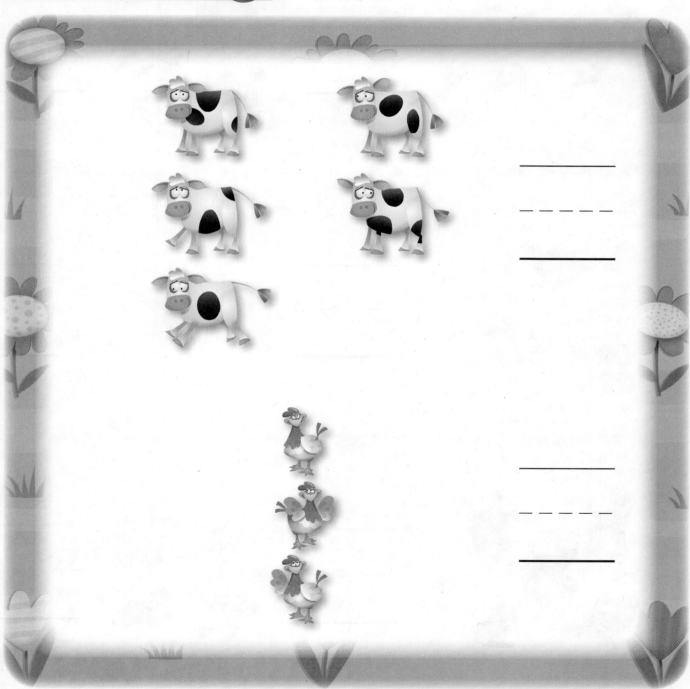

DIRECTIONS Look at the sets of objects. Count how many objects in each set. Write the numbers. Compare the numbers and tell a friend which number is greater and which number is less.

Chapter 2 • Lesson 5

one hundred five **105**

1

_____

_____

2

_____

_____

3 ✓

_____

_____

DIRECTIONS 1–3. Count how many objects in each set.
Write the numbers. Compare the numbers. Circle the number
that is greater.

Name _____

_____

- - - - - - - - -

_____

_____

- - - - - - - - -

_____

❋ 5

_____

- - - - - - - - -

_____

_____

- - - - - - - - -

_____

🌰 6

_____

- - - - - - - - -

_____

_____

- - - - - - - - -

_____

DIRECTIONS   4–6. Count how many objects in each set.
Write the numbers. Compare the numbers. Circle the number
that is less.

Problem Solving • Applications (Real World)

7

_____

- - - - - - - - -

_____

- - - - - - - - -

_____

8

_____

- - - - - - - - -

_____

_____

- - - - - - - - -

_____

DIRECTIONS 7. Tony has stuffed toy frogs. His friend has stuffed toy turkeys. Count how many objects in each set. Write the numbers. Compare the numbers. Tell a friend what you know about the sets. **8.** Draw to show what you know about counting to compare two sets of objects. Write how many in each set.

HOME ACTIVITY • Draw a domino block with up to three dots on one end. Ask your child to draw on the other end a number of dots greater than the set you drew.

108 one hundred eight

Name _____

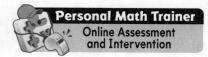

_____

- - - - - - - -

_____

_____

- - - - - - - -

_____

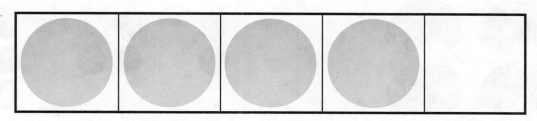

_____

- - - - - - - -

_____

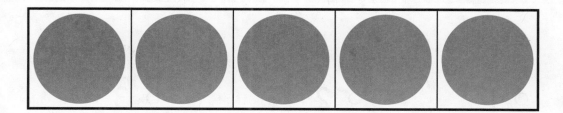

_____

- - - - - - - -

_____

DIRECTIONS 1. Draw a counter below each finger puppet to show the same number of counters as puppets. Write how many puppets. Write how many counters. 2. How many counters are there in each row? Write the numbers. Compare the sets by matching. Circle the number that is greater.

Chapter 2

 Assessment Options
Chapter Test

one hundred eleven **111**

3

○   ○   ○

4

○   ○   ○

5

○   ○   ○

6

○ 1   ○ 2   ○ 3

DIRECTIONS   **3.** Mark under all the sets that have the same number of counters as the number of cars.   **4.** Mark under all the sets that have a number of counters greater than the number of turtles.   **5.** Mark under all the sets that have a number of counters less than the number of vans.   **6.** Mark all the numbers less than 3.

112   one hundred twelve

Name _____

 7

_____
- - - - - - -
_____

_____
- - - - - - -
_____

Personal Math Trainer

8 THINK SMARTER +

_____
- - - - - - -
_____

_____
- - - - - - -
_____

DIRECTIONS **7.** Maria has these apples. Draw a set of oranges below the apples that has the same number. Compare the sets by matching. Write how many pieces of fruit in each set. **8.** Amy has two crayons. Draw Amy's crayons. Brad has I more crayon than Amy. How many crayons does Brad have? Draw Brad's crayons. Write how many in each set.

Chapter 2

9 THINK SMARTER +

• same number

• greater than

• less than

10

_____

- - - - -

_____

_____

- - - - -

_____

DIRECTIONS **9.** Compare the number of red counters in each set to the number of blue counters. Draw lines from the sets of counters to the words that show *same number*, *greater than*, or *less than*. **10.** Draw four counters. Now draw a set that has a greater number of counters. How many are in each set? Write the numbers. Use green to color the set with a greater number of counters. Use blue to color the set with a number of counters that is less than the green set.

114 one hundred fourteen